Table of Contents

Unit 1 .... **3**
- Numerals (1–10)
- Number words (one to ten)

Unit 2 .... **17**
- Shapes
    - circle
    - square
    - triangle
    - rectangle
    - oval
    - diamond
    - star
- Shapes in a picture
- Shape mazes
- Matching shapes
- Matching shapes with objects

Unit 3 .... **44**
- Tracing
    - shapes
    - shapes to complete a picture
- Fine motor skills (mazes)

Unit 4 .... **61**
- Counting to 10
- Number sequence to 10

Unit 5 .... **89**
- Sets
    - with the same number
    - with more
    - with less
- Counting objects in a set
- Matching numerals to sets

Unit 6 ....
- Visual discrimination
    - same
    - different
- Sizes
    - same size
    - tall
    - short (height)
    - big
    - little
    - long
    - short (length)
    - size differentiation
        - tallest
        - shortest (height)
        - biggest
        - littlest
        - longest
        - shortest (length)
- Shape patterns
- Number patterns

Unit 7 .... **135**
- Time
    - things that tell time
    - length of time
- Money
    - identifying money
    - amount of money
- Measurement (things that measure)

Unit 8 .... **143**
Comprehensive skills practice test in fill-in-the-circle format.

Answer Key **151**

How to Use This Book

This Premium Education Series workbook is designed to suit your teaching needs. Since every child learns at his or her own pace, this workbook can be used individually or as part of small group instruction. The activity pages can be used together with other educational materials and are easily applied to a variety of teaching approaches.

Contents
A detailed table of contents lists all the skills that are covered in the workbook.

Units
The workbook is divided into units of related skills. Numbered tabs allow you to quickly locate each unit. The skills within each unit are designed to be progressively more challenging.

Activity Pages
Each activity page is titled with the skill being practiced or reinforced. The activities and units in this workbook can be used in sequential order, or they can be used to accommodate and supplement any educational curriculum. In addition, the activity pages include simple instructions to encourage independent study, and they are printed in black and white so they can be easily reproduced. Plus, you can record the child's name and the date the activity was completed on each page to keep track of learning progress.

Teaching Tips
Some of the activity pages include teaching tips, which are designed to help you get the most out of the activity. They can also help you extend the learning experience beyond the workbook page.

Practice Test
A comprehensive practice test helps prepare the child for standardized testing in a stress-free environment. Standardized testing can be a part of school curriculum as early as kindergarten, so it is important for a child to feel confident with the fill-in-the-circle testing format.

Answer Key
The pages in the back of the workbook provide answers for each activity page as well as the practice test. These answer pages allow you to quickly check the child's work and provide immediate feedback on how he or she is progressing.

Numerals: 1 (one)

Name_____ Date_____

1 one

Unit 1

Color the **one** hippo.

Trace and print the numeral.

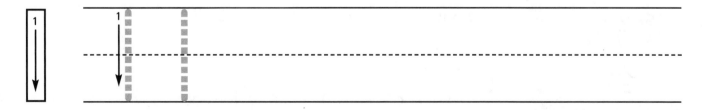

Teaching Tip: Ask the child point to the numeral and number word at the top and help him or her say "one." Have the child color the hippo. At the bottom, encourage the child to use a finger to trace the numeral before practicing writing with a crayon or pencil.

Premium Education Math: Preschool © Learning Horizons

Numerals: 2 (two)

Name_____ Date_____

2 two

Color the **two** clowns.

Trace and print the numeral.

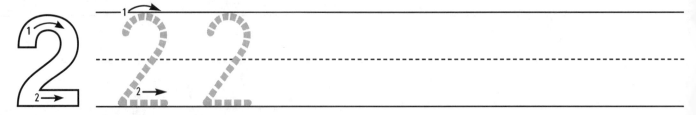

Teaching Tip: Ask the child point to the numeral and number word at the top and help him or her say "two." Have the child color the clowns. At the bottom, encourage the child to use a finger to trace the numeral before practicing writing with a crayon or pencil.

Premium Education Math: Preschool © Learning Horizons

Numerals: 3 (three)

Name_____ Date_____

3 three

Unit 1

Color the **three** jack-in-the-boxes.

Trace and print the numeral.

Teaching Tip: Ask the child point to the numeral and number word at the top and help him or her say "three." Have the child color the jack-in-the-boxes. At the bottom, encourage the child to use a finger to trace the numeral before practicing writing with a crayon or pencil.

Premium Education Math: Preschool © Learning Horizons

Numerals: 4 (four)

Name_____   Date_____

4 four

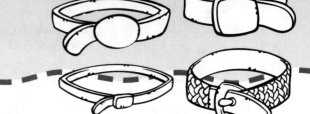

Color the **four** sweaters.

Trace and print the numeral.

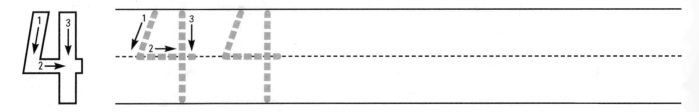

Teaching Tip: Ask the child point to the numeral and number word at the top and help him or her say "four." Have the child color the sweaters. At the bottom, encourage the child to use a finger to trace the numeral before practicing writing with a crayon or pencil.

Premium Education Math: Preschool © Learning Horizons

Numerals: 5 (five)

Name_____ Date_____

5 five

Unit 1

Color the **five** jack-o-lanterns.

Trace and print the numeral.

 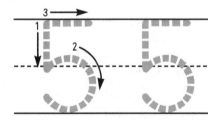

Teaching Tip: Ask the child point to the numeral and number word at the top and help him or her say "five." Have the child color the jack-o-lanterns. At the bottom, encourage the child to use a finger to trace the numeral before practicing writing with a crayon or pencil.

Premium Education Math: Preschool   © Learning Horizons

Numerals: 1-5 Review

Name_____ Date_____

Trace the numerals.

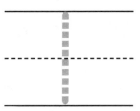

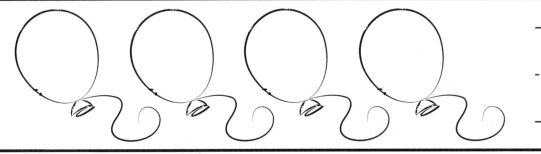
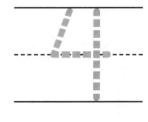

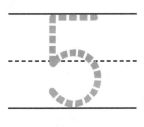

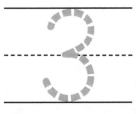

Numerals: One to Five Review

Name_____ Date_____

Unit 1

Match the numerals and number words with the pictures.

1 one

3 three

2 two

5 five

4 four

Numerals: 6 (six)

Name_____ Date_____

6 six

Color the six dogs.

Trace and print the numeral.

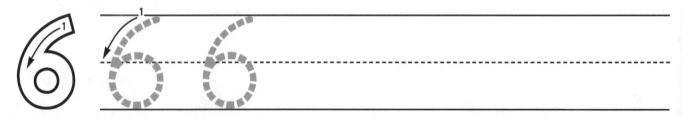

Teaching Tip: Ask the child point to the numeral and number word at the top and help him or her say "six." Have the child color the dogs. At the bottom, encourage the child to use a finger to trace the numeral before practicing writing with a crayon or pencil.

Numerals: 7 (seven)

Name_____ Date_____

7 seven

Unit 1

Color the **seven** grasshoppers.

Trace and print the numeral.

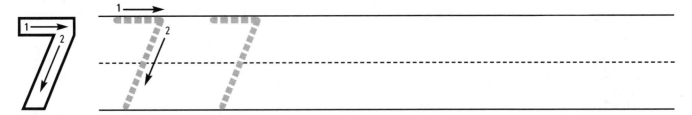

Teaching Tip: Ask the child point to the numeral and number word at the top and help him or her say "seven." Have the child color the grasshoppers. At the bottom, encourage the child to use a finger to trace the numeral before practicing writing with a crayon or pencil.

Numerals: 8 (eight)

Name_____ Date_____

8 eight

Color the **eight** strawberries.

Trace and print the numeral.

 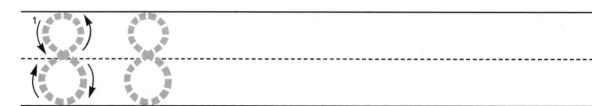

Teaching Tip: Ask the child point to the numeral and number word at the top and help him or her say "eight." Have the child color the strawberries. At the bottom, encourage the child to use a finger to trace the numeral before practicing writing with a crayon or pencil.

Numerals: 9 (nine)

Name_____ Date_____

9 nine

Unit 1

Color the **nine** butterflies.

Trace and print the numeral.

 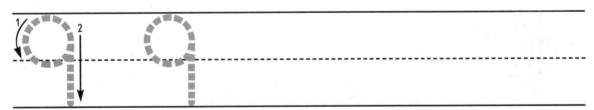

Teaching Tip: Ask the child point to the numeral and number word at the top and help him or her say "nine." Have the child color the butterflies. At the bottom, encourage the child to use a finger to trace the numeral before practicing writing with a crayon or pencil.

Premium Education Math: Preschool   13   © Learning Horizons

Numerals: 10 (ten)

Name_____ Date_____

10 ten

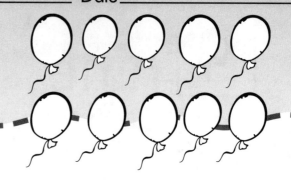

Color the **ten** cupcakes.

Trace and print the numeral.

 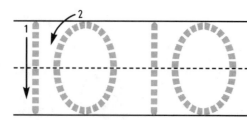

Teaching Tip: Ask the child point to the numeral and number word at the top and help him or her say "ten." Have the child color the cupcakes. At the bottom, encourage the child to use a finger to trace the numeral before practicing writing with a crayon or pencil.

Premium Education Math: Preschool   © Learning Horizons

Numerals: Six to Ten Review

Name_____ Date_____

Unit 1

Match the numerals and number words with the pictures.

6 six

9 nine

8 eight

10 ten

7 seven

Numerals: 1-10 Review

Name_____ Date_____

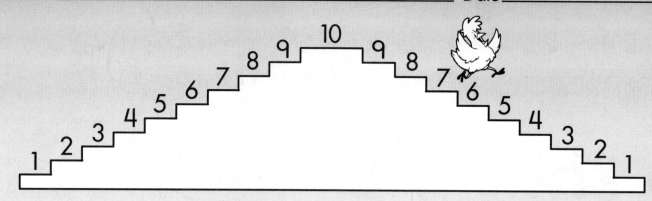

Trace and write the numerals **1** to **10**.

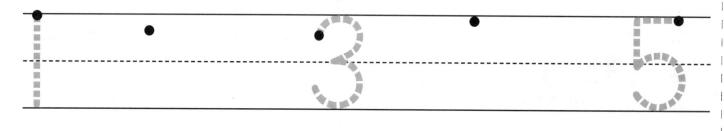

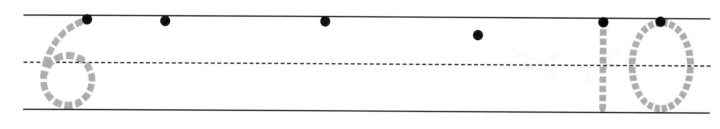

Trace and write the numerals **10** to **1**.

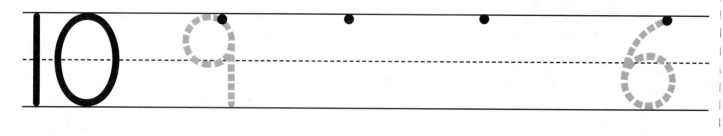

Premium Education Math: Preschool © Learning Horizons

Shapes: Circle (1)

Name_____ Date_____

Unit 2

Color the ◯s.

Shapes: Circle (II)

Name_____ Date_____

Find and color the ◯s.

Shapes: Circle (III)

Name_____ Date_____

Color the ◯s to make a path to the gingerbread house.

Unit 2

Shapes: Square (1)

Name_____ Date_____

Color the ☐ s.

Shapes: Square (II)

Name_____ Date_____

Find and color the ▢s.

Unit 2

Shapes: Square (III)

Name_____ Date_____

Color the □s to make a path to the doghouse.

Shapes: Triangle (I)

Name_____ Date_____

Unit 2

Color the △s.

Shapes: Triangle (II)

Name_____ Date_____

Find and color the △s.

Shapes: Triangle (III)

Name_____ Date_____

Color the △s to make a path to the nest.

Unit 2

Shapes: Rectangle (1)

Name_____ Date_____

Color the ▭s.

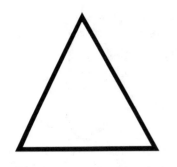 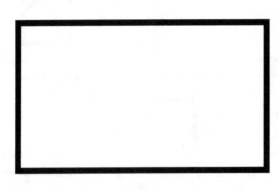 

Shapes: Rectangle (II)

Name_____ Date_____

Find and color the ▭s.

Unit 2

Shapes: Rectangle (III)

Name_____ Date_____

Color the ▭s to make a path to the mother duck.

Shapes: Oval (I)

Name_____ Date_____

Unit 2

Color the ◯s.

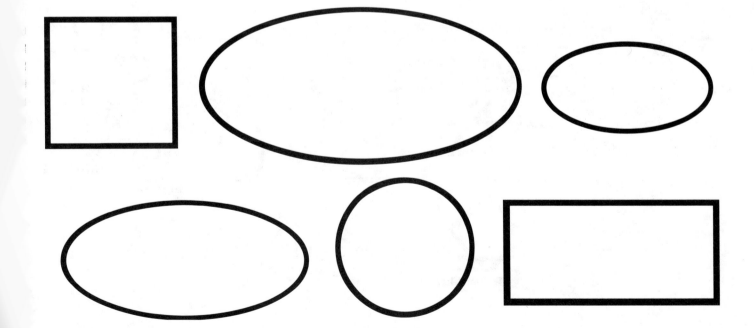

Shapes: Oval (II)

Name_____ Date_____

Find and color the ⬭s.

Shapes: Oval (III)

Name_____ Date_____

Color the ⬭s to make a path to the finish line.

Unit 2

Shapes: Diamond (1)

Name_____ Date_____

Color the ◇s.

Shapes: Diamond (II)

Name_____ Date_____

Find and color the ◇s.

Unit 2

Shapes: Diamond (III)

Name_____ Date_____

Color the ◇s to make a path to the yarn.

Premium Education Math: Preschool © Learning Horizons

Shapes: Star (1)

Name_____ Date_____

Unit 2

Color the ☆s.

Shapes: Star (II)

Name_____ Date_____

Find and color the ☆s.

Shapes: Star (III)

Name_____ Date_____

Color the ☆s to make a path to the carrot.

Unit 2

Shapes: Matching (1)

Name_____ Date_____

Circle the shapes that are the same in each row.

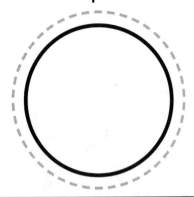  

  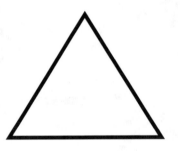

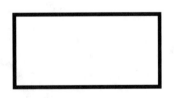 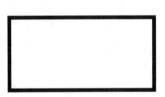 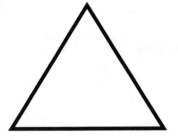

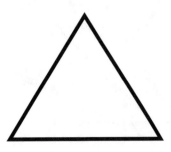 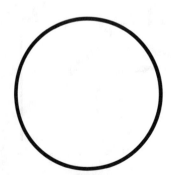 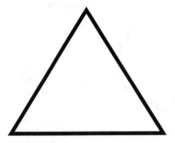

Shapes: Matching (II)

Name_____ Date_____

Color the shapes that are the same in each row.

 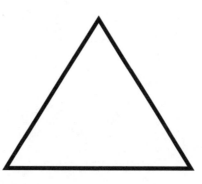 

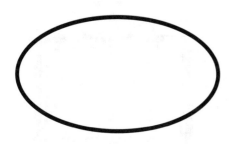 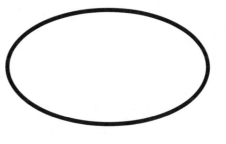 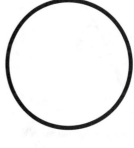

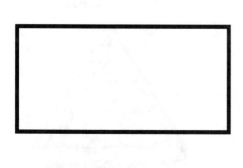 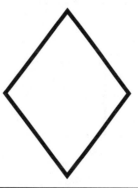 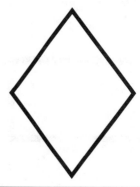

  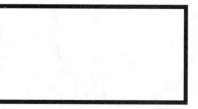

Unit 2

Shapes: Matching (III)

Name_____ Date_____

Match the shapes.

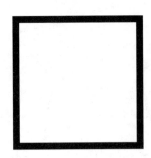

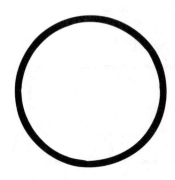

Shapes: Matching (IV)

Name_____   Date_____

Match the shapes.

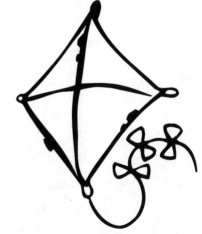

Unit 2

Shapes: Review (1)

Name_____ Date_____

Use the color key to color the picture.

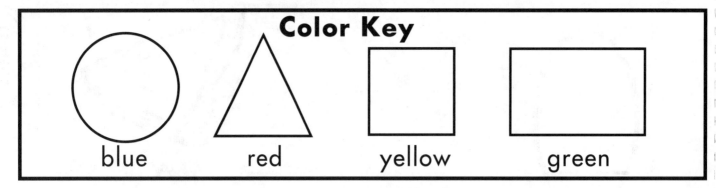

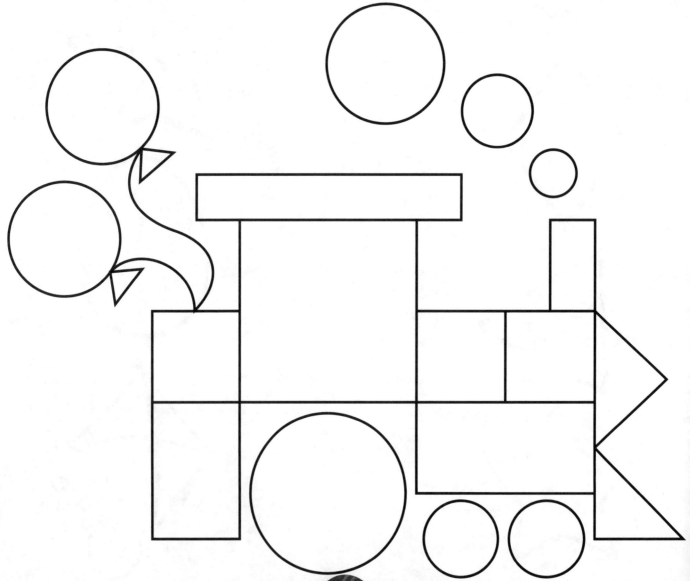

Shapes: Review (II)

Name_____ Date_____

Use the color key to color the picture.

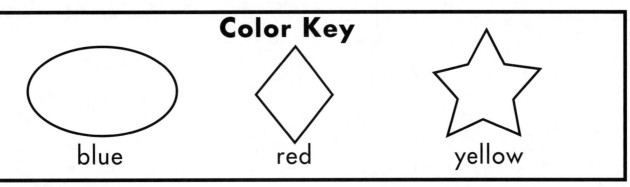

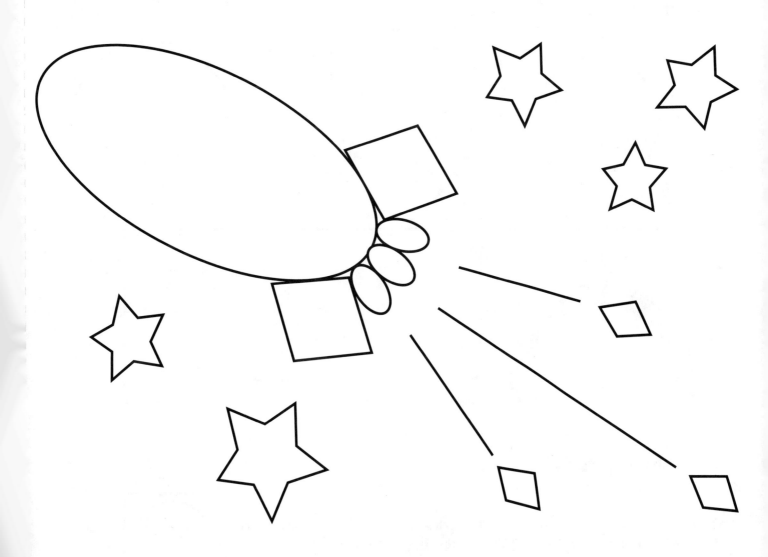

Unit 2

Tracing: Circles (1)

Name_____ Date_____

Color the circles.

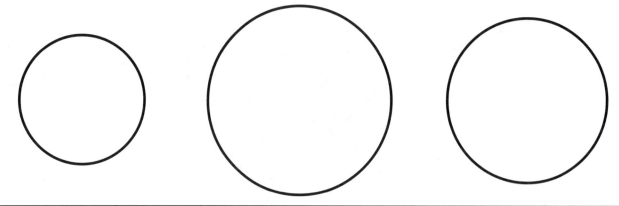

Trace and draw the circles.

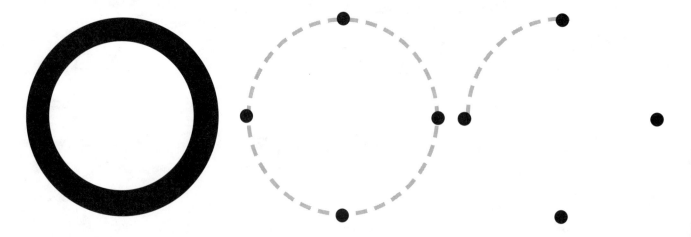

Teaching Tip: Ask the child to color the circles at the top. At the bottom, encourage him or her to use a finger to trace the bolded shape before practicing drawing with a crayon or pencil.

Premium Education Math: Preschool   © Learning Horizons

Tracing: Circles (II)

Name_____ Date_____

Trace the circles to complete the picture. Color the picture.

Unit 3

Teaching Tip: Encourage the child to use a finger to trace the shapes before drawing with a crayon or pencil.

Tracing: Squares (I)

Name_____ Date_____

Color the squares.

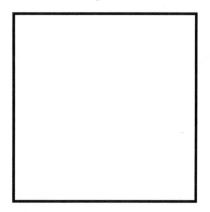 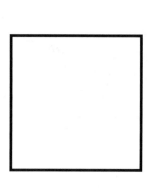 

Trace the squares.

 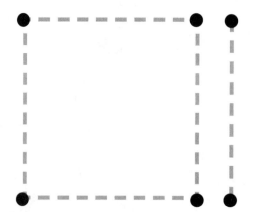

Teaching Tip: Ask the child to color the squares at the top. At the bottom, encourage him or her to use a finger to trace the bolded shape before practicing drawing with a crayon or pencil.

Tracing: Squares (II)

Name_____ Date_____

Trace the squares to complete the picture. Color the picture.

Unit 3

Teaching Tip: Encourage the child to use a finger to trace the shapes before drawing with a crayon or pencil.

Premium Education Math: Preschool © Learning Horizons

Tracing: Triangles (1)

Name_____ Date_____

Color the triangles.

Trace and draw the triangles.

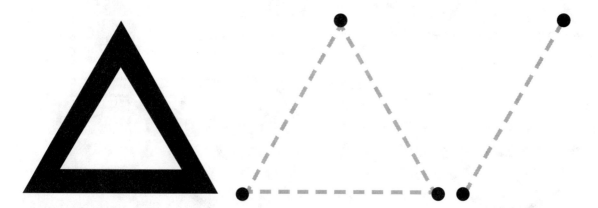

Teaching Tip: Ask the child to color the triangles at the top. At the bottom, encourage him or her to use a finger to trace the bolded shape before practicing drawing with a crayon or pencil.

Tracing: Triangles (II)

Name_____ Date_____

Trace the triangles to complete the picture. Color the picture.

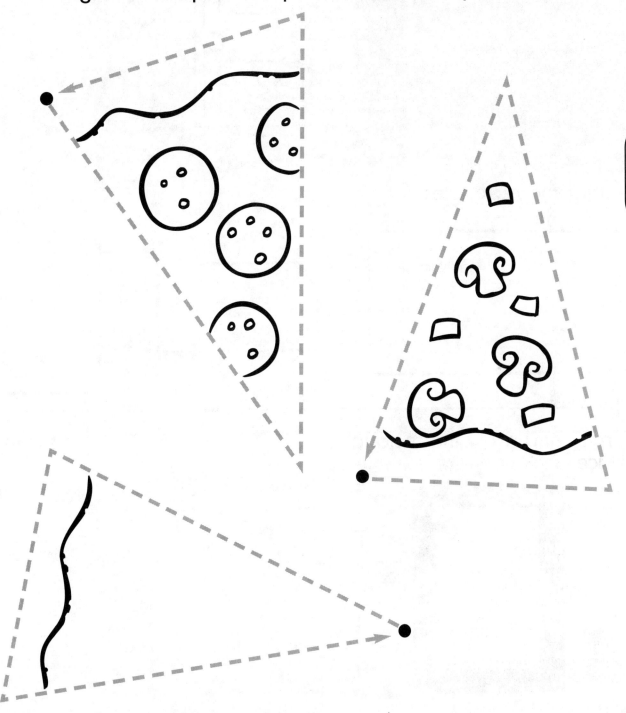

Teaching Tip: Encourage the child to use a finger to trace the shapes before drawing with a crayon or pencil.

Tracing: Rectangles (1)

Name_____ Date_____

Color the rectangles.

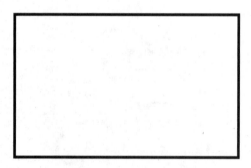 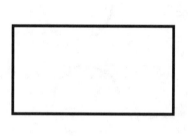 

Trace and draw the rectangles.

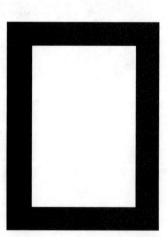 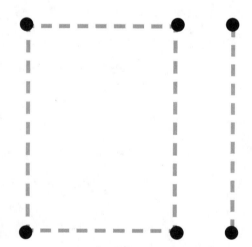

Teaching Tip: Ask the child to color the rectangles at the top. At the bottom, encourage him or her to use a finger to trace the bolded shape before practicing drawing with a crayon or pencil.

Tracing: Rectangles (II)

Name_____ Date_____

Trace the rectangles to complete the picture. Color the picture.

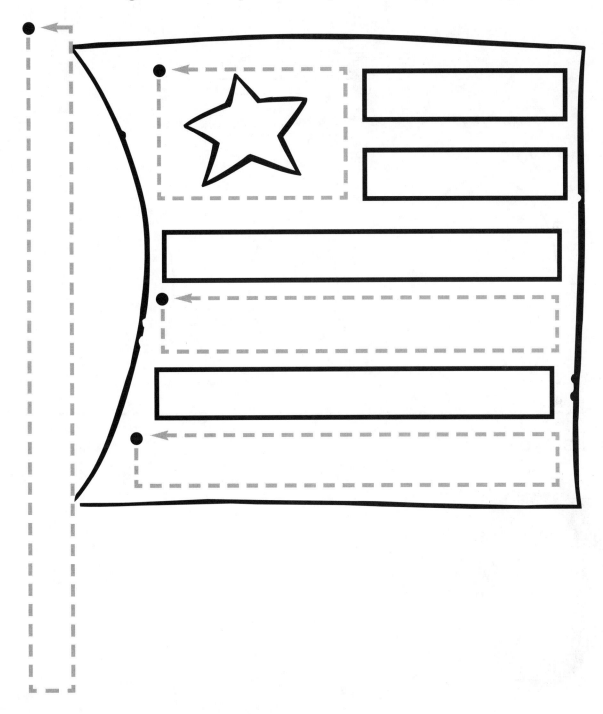

Teaching Tip: Encourage the child to use a finger to trace the shapes before drawing with a crayon or pencil.

Tracing: Ovals (1)

Name_____ Date_____

Color the ovals.

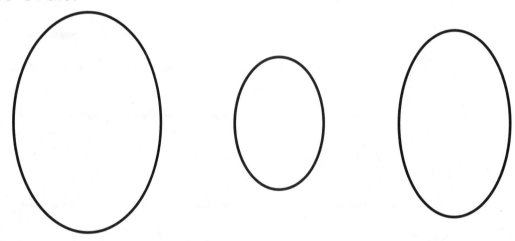

Trace the ovals.

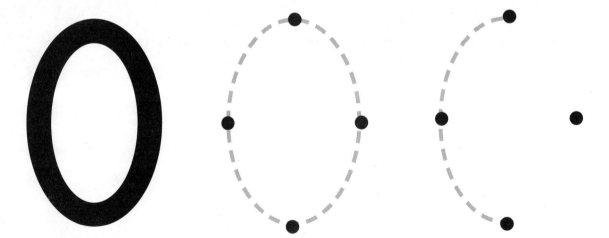

Teaching Tip: Ask the child to color the ovals at the top. At the bottom, encourage him or her to use a finger to trace the bolded shape before practicing drawing with a crayon or pencil.

Tracing: Ovals (II)

Name_____ Date_____

Trace the ovals to complete the picture. Color the picture.

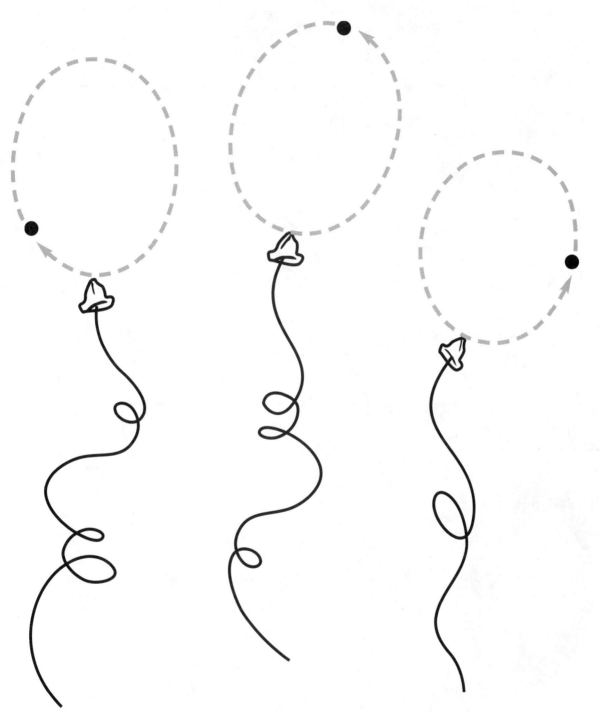

Teaching Tip: Encourage the child to use a finger to trace the shapes before drawing with a crayon or pencil.

Tracing: Diamonds (1)

Name_____ Date_____

Color the diamonds.

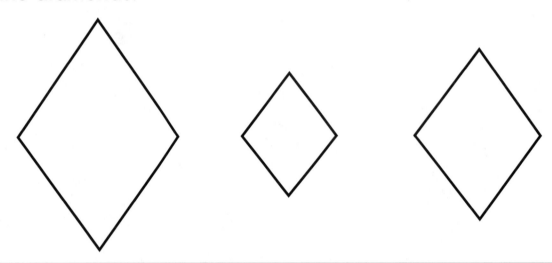

Trace the diamonds.

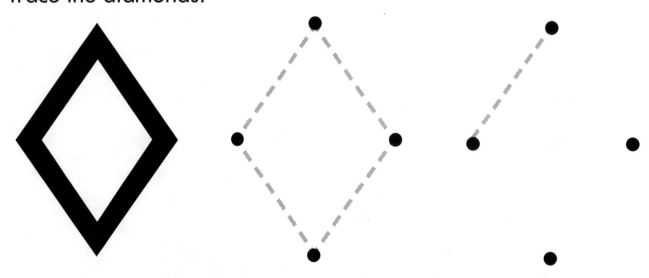

Teaching Tip: Ask the child to color the diamonds at the top. At the bottom, encourage him or her to use a finger to trace the bolded shape before practicing drawing with a crayon or pencil.

Tracing: Diamonds (II)

Name_____ Date_____

Trace the diamonds to complete the picture. Color the picture.

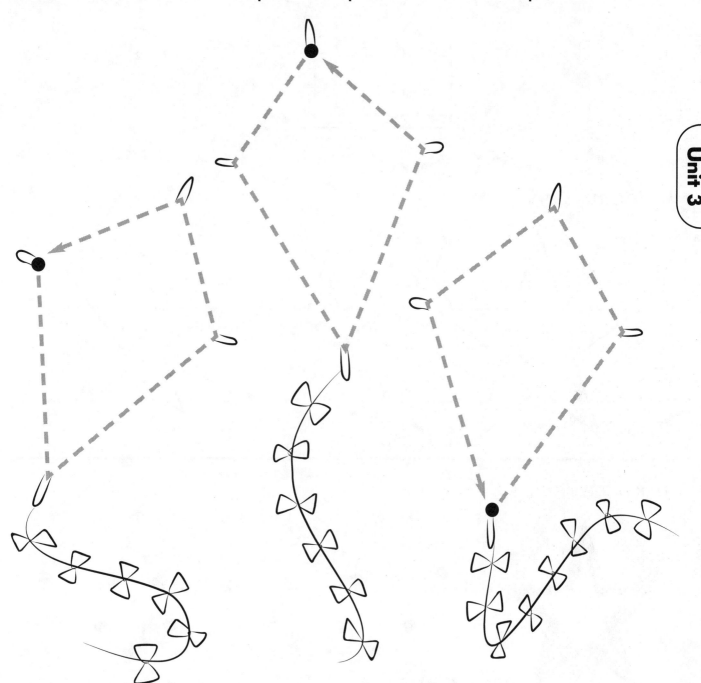

Teaching Tip: Encourage the child to use a finger to trace the shapes before drawing with a crayon or pencil.

Premium Education Math: Preschool © Learning Horizons

Tracing: Stars (1)

Name_____ Date_____

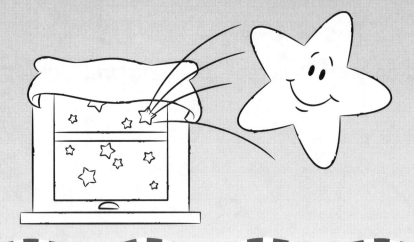

Color the stars.

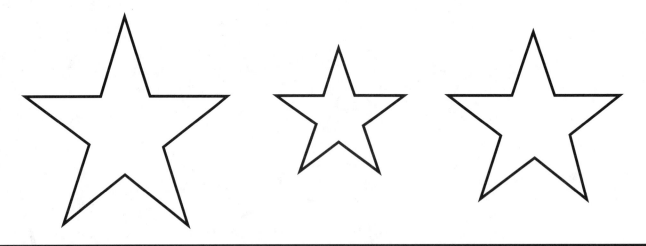

Trace the stars.

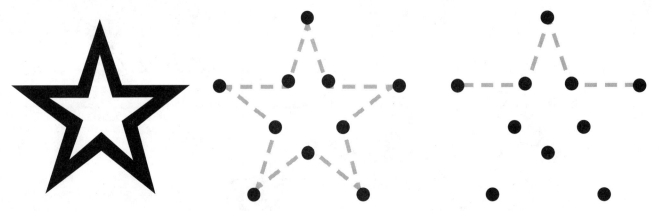

Teaching Tip: Ask the child to color the stars at the top. At the bottom, encourage him or her to use a finger to trace the bolded shape before practicing drawing with a crayon or pencil.

Tracing: Stars (II)

Name_____ Date_____

Trace the stars to complete the picture. Color the picture.

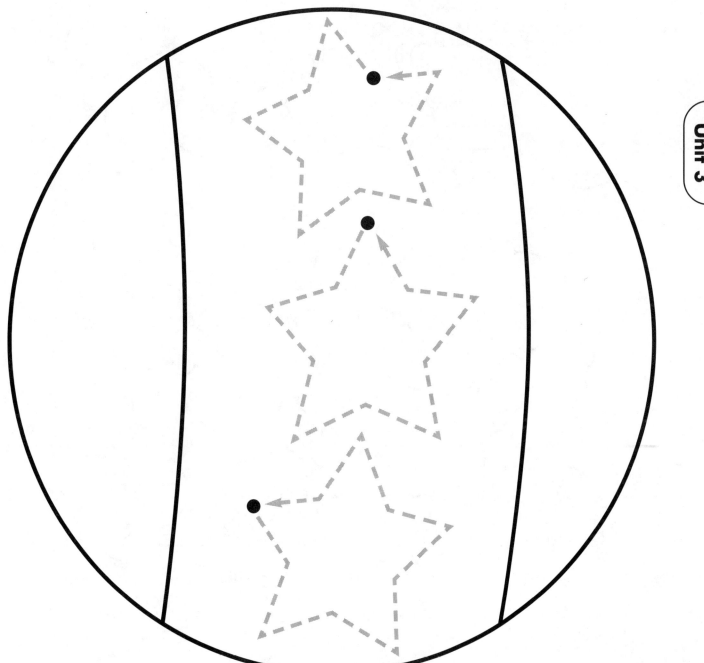

Unit 3

Teaching Tip: Encourage the child to use a finger to trace the shapes before drawing with a crayon or pencil.

Fine Motor Skills: Mazes (I)

Name_____ Date_____

Draw the path to the bird's nest.

Teaching Tip: Encourage the child to use his or her finger to find the way through the maze before drawing the path with a crayon.

Fine Motor Skills: Mazes (II)

Name_____ Date_____

Draw the path to the treasure chest.

Teaching Tip: Encourage the child to use his or her finger to find the way through the maze before drawing the path with a crayon.

Fine Motor Skills: Mazes (III)

Name_____ Date_____

Draw the path to the cheese.

Teaching Tip: Encourage the child to use his or her finger to find the way through the maze before drawing the path with a crayon.

Counting 1 (1)

Name_____ Date_____

Count and color **1**.

Count and circle **1**.

Counting 1 (II)

Name_____ Date_____

Count and color **1** in each group.

Counting 2 (1)

Name_____ Date_____

Count and color **2**.

Count and circle **2**.

Counting 2 (II)

Name_____ Date_____

Count and color **2** in each group.

Counting 3 (1)

Name_____ Date_____

Count and color **3**.

Count and circle **3**.

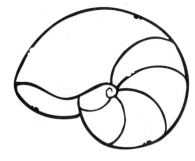
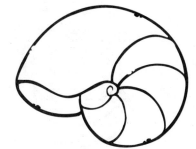
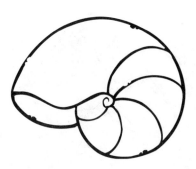
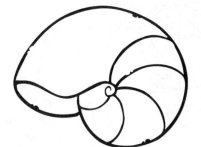
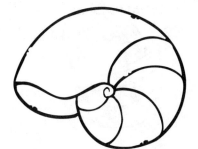

Unit 4

Counting 3 (II)

Name_____ Date_____

Count and color **3** in each group.

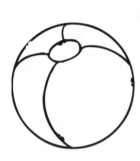 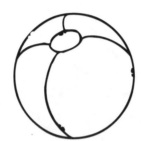  

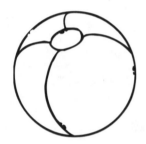 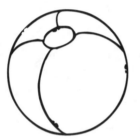 

Counting 4 (1)

Name_____ Date_____

Count and color 4.

Count and circle 4.

Unit 4

Counting 4 (II)

Name_____ Date_____

Count and color **4** in each group.

Counting 5 (I)

Name_____ Date_____

Count and color **5**.

Unit 4

Count and circle **5**.

Counting 5 (II)

Name_____ Date_____

Count and color **5** in each group.

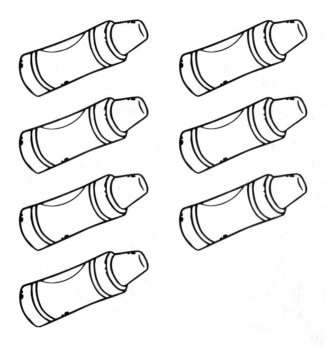
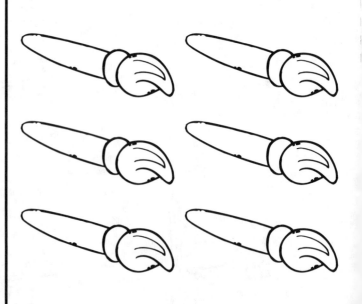

Number Sequence to 5: Dot-to-Dots (1)

Name_____ Date_____

Connect the dots from **1** to **5** to complete the picture. Color the picture.

Number Sequence to 5: Dot-to-Dots (II)

Name_____ Date_____

Connect the dots from **1** to **5** to complete the picture. Color the picture.

Counting to 5: Review (1)

Name_____ Date_____

Draw a line to show how many.

1

2

3

4

5

Unit 4

Counting to 5: Review (II)

Name_____ Date_____

Circle the correct number.

3   (4)   5   3   4   5

3   4   5   1   2   3

Counting 6 (1)

Name_____ Date_____

Count and color **6**.

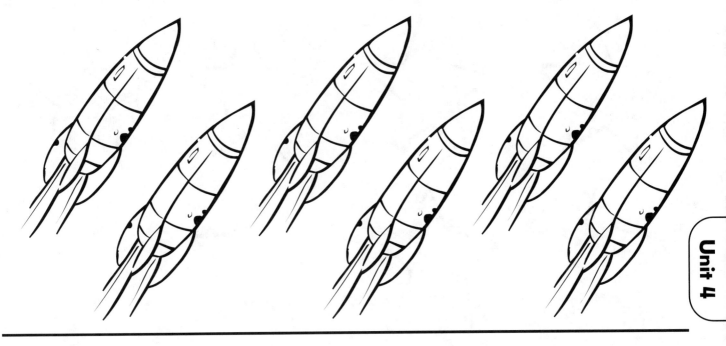

Count and circle **6**.

Unit 4

Counting 6 (II)

Name_____ Date_____

Count and color **6** in each group.

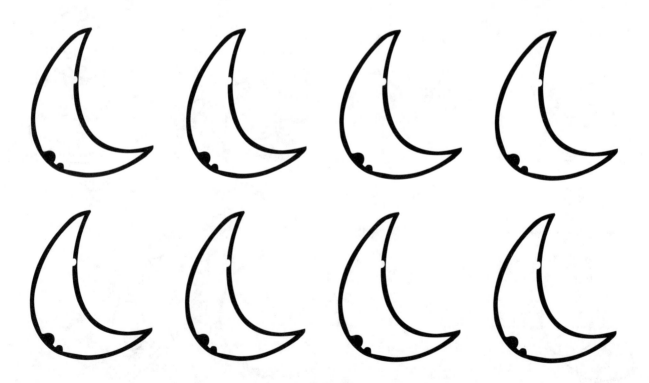

Counting 7 (I)

Name_____ Date_____

Count and color **7**.

Count and circle **7**.

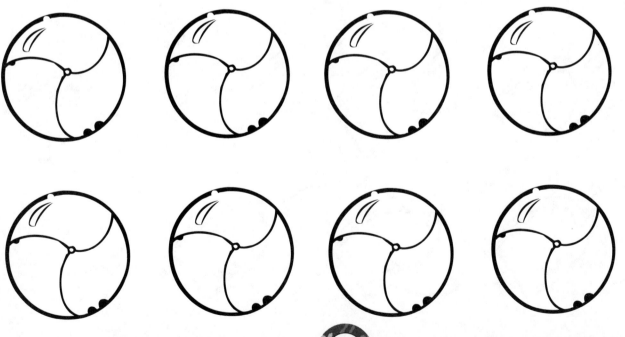

Unit 4

Counting 7 (II)

Name_____ Date_____

Count and color **7** in each group.

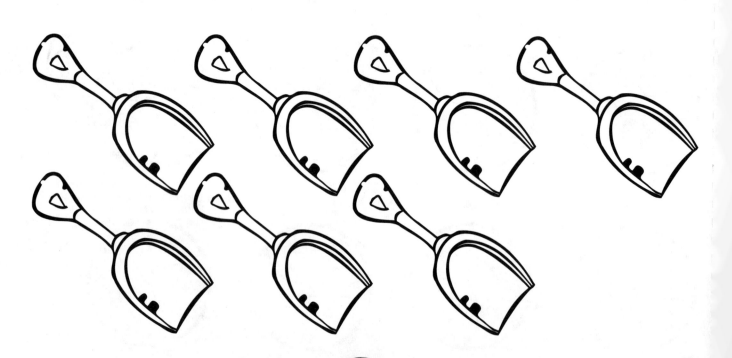

Counting 8 (I)

Name_____ Date_____

Count and color **8**.

Count and circle **8**.

Counting 8 (II)

Name_____ Date_____

Count and color **8** in each group.

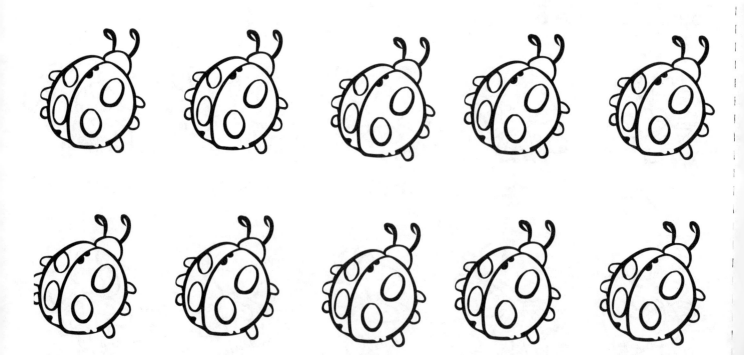

Counting 9 (1)

Name_____ Date_____

Count and color 9.

Count and circle 9.

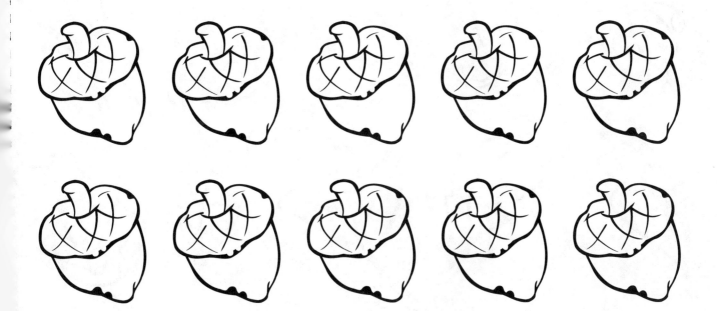

Unit 4

Counting 9 (II)

Name_____ Date_____

Count and color **9** in each group.

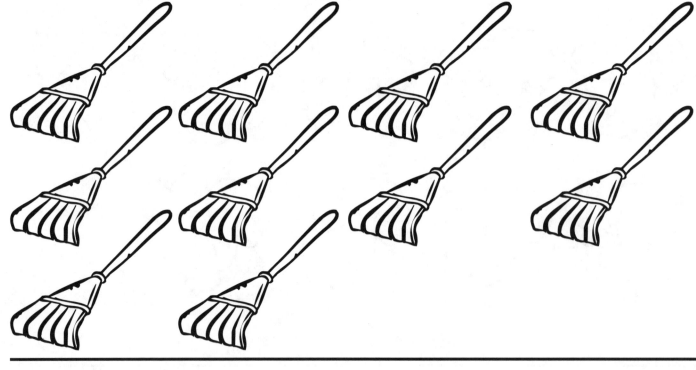

Counting 10 (I)

Name_____ Date_____

Count and color **10**.

Unit 4

Count and circle **10**.

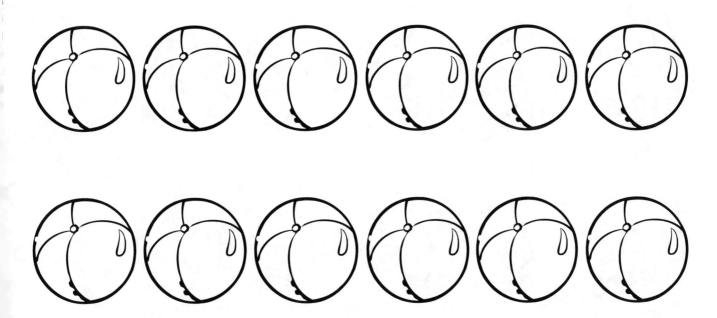

Counting 10 (II)

Name_____ Date_____

Count and color **10** in each group.

Number Sequence to 10: Dot-to-Dots (1)

Name_____ Date_____

Connect the dots from **1** to **10** to finish the picture. Color the picture.

Number Sequence to 10: Dot-to-Dots (II)

Name_____ Date_____

Connect the dots from **1** to **10** to finish the picture. Color the picture.

Counting to 10: Review (1)

Name_____ Date_____

Circle the correct number.

1 **2** **3**

6 **7** **8**

8 **9** **10**

4 **5** **6**

2 **3** **4**

7 **8** **9**

Unit 4

Counting to 10: Review (II)

Name_____ Date_____

Color the correct number of pictures.

4

9

5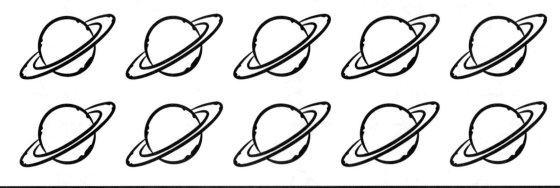

6

Sets: Matching (1)

Name_____ Date_____

Match the ones with the same number.

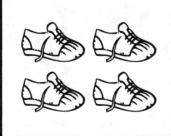

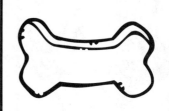
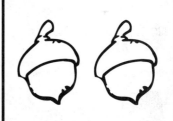
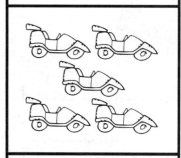
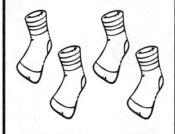

Unit 5

Sets: Matching (II)

Name_____ Date_____

Match the ones with the same number.

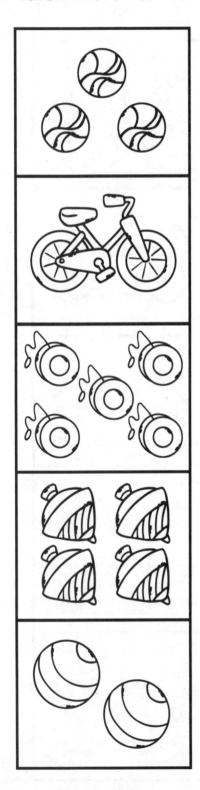
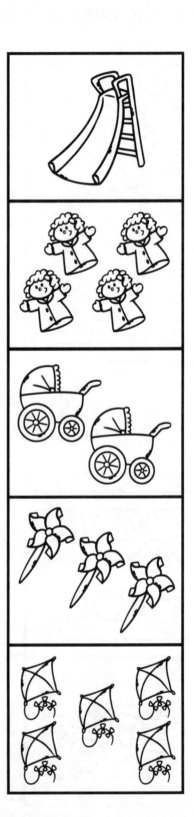

Sets: Matching (III)

Name_____ Date_____

Match the ones with the same number.

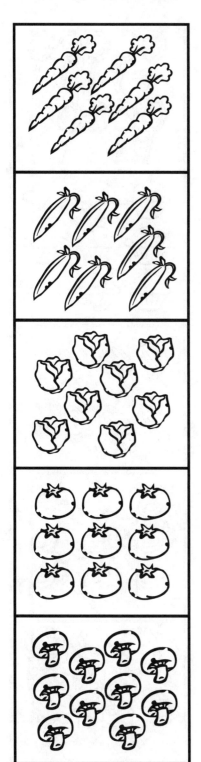
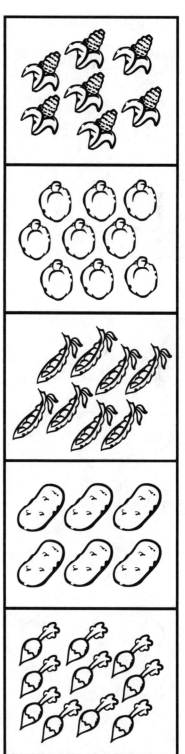

Sets: Matching (IV)

Name_____ Date_____

Match the ones with the same number.

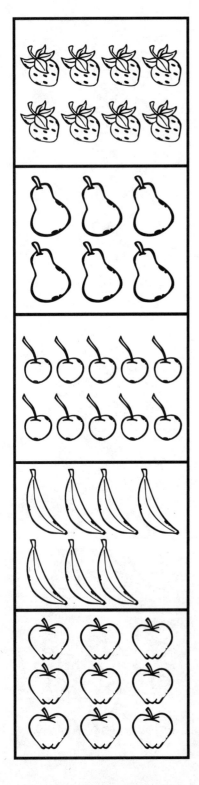

Sets: More (1)

Name_____ Date_____

Circle the group with **more**.

Teaching Tip: Ask the child to identify the pictures on the page. Then have him or her count how many are in each group. Help the child understand which group has more.

Sets: More (II)

Name_____ Date_____

Color the one with **more**.

Teaching Tip: Ask the child to identify the pictures on the page. Then have him or her count how many are in each group. Help the child understand which group has more.

Sets: More (III)

Name_____ Date_____

Circle the group with **more**.

 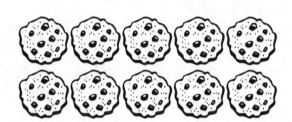

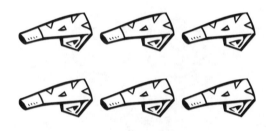 

Teaching Tip: Ask the child to identify the pictures on the page. Then have him or her count how many are in each group. Help the child understand which group has more.

Sets: Less (1)

Name_____ Date_____

Circle the group with **less**.

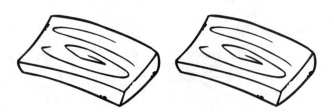 

 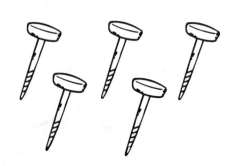

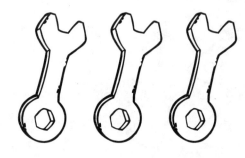 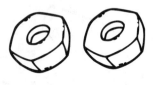

 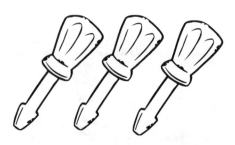

Teaching Tip: Ask the child to identify the pictures on the page. Then have him or her count how many are in each group. Help the child understand which group has less.

Sets: Less (II)

Name_____ Date_____

Circle the group with **less**.

 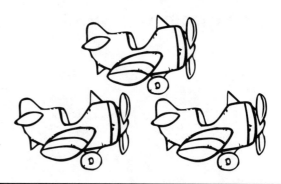

Teaching Tip: Ask the child to identify the pictures on the page. Then have him or her count how many are in each group. Help the child understand which group has less.

Sets: Less (III)

Name_____ Date_____

Circle the group with **less**.

 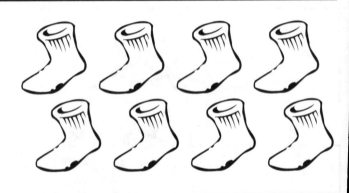

Teaching Tip: Ask the child to identify the pictures on the page. Then have him or her count how many are in each group. Help the child understand which group has less.

Counting 1-5 (I)

Name_____ Date_____

Count and write the number.

Unit 5

Counting 6-10 (II)

Name_____ Date_____

Count and write the number.

 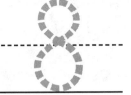

 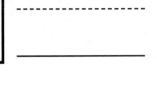

Counting Objects in a Set (1)

Name_____   Date_____

Count and write the number.

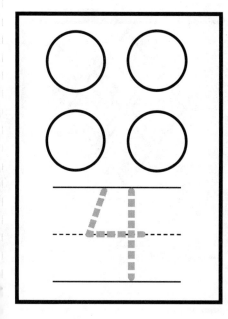

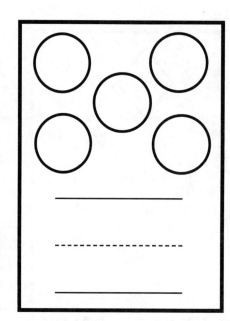
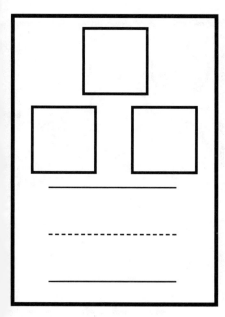
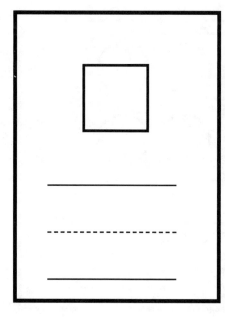
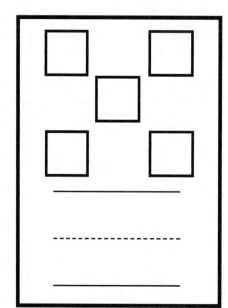

Unit 5

Counting Objects in a Set (II)

Name _____   Date _____

Count and write the number.

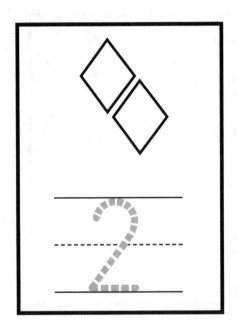

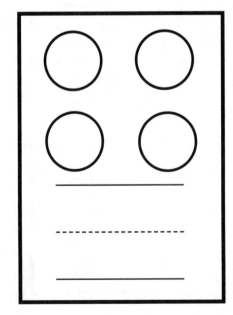
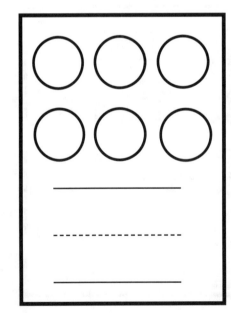
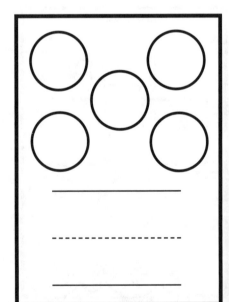

Matching Sets to Numerals (1)

Name_____ Date_____

Match the group to the number.

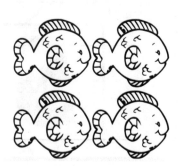 

Unit 5

Teaching Tip: Encourage the child to use a finger to count the fish in each group. Then have him or her slide the finger to the numeral that shows how many before drawing the line with a crayon or pencil.

Premium Education Math: Preschool © Learning Horizons

Matching Sets to Numerals (II)

Name_____ Date_____

Match the group to the number.

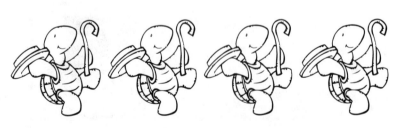

Teaching Tip: Encourage the child to use a finger to count the animals in each group. Then have him or her slide the finger to the numeral that shows how many before drawing the line with a crayon or pencil.

Premium Education Math: Preschool © Learning Horizons

Matching Sets to Numerals (III)

Name_____ Date_____

Match the group to the number.

Unit 5

Teaching Tip: Encourage the child to use a finger to count the fruit in each group. Then have him or her slide the finger to the numeral that shows how many before drawing the line with a crayon or pencil.

Matching Sets to Numerals (IV)

Name_____ Date_____

Match the group to the number.

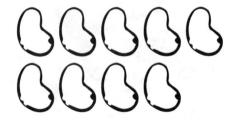

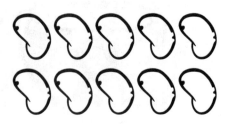

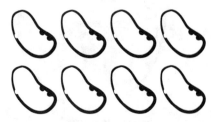

Teaching Tip: Encourage the child to use a finger to count the jelly beans in each group. Then have him or her slide the finger to the numeral that shows how many before drawing the line with a crayon or pencil.

Visual Discrimination: Same (I)

Name_____ Date_____

Color the two that are the **same**.

Unit 6

Teaching Tip: Have the child look closely at the toys in each group. Then have him or her point to the ones that are the same before circling them. Encourage the child to tell you how they are the same.

Visual Discrimination: Same (II)

Name_____ Date_____

Color the two that are the **same**.

Visual Discrimination: Same (III)

Name_____ Date_____

Circle the two that are the **same**.

Visual Discrimination: Different (I)

Name_____ Date_____

Circle the one that is **different**.

Teaching Tip: Have the child look closely at the animals in each group. Then have him or her point to the one that is different before circling it. Encourage the child to tell you how it is different.

Visual Discrimination: Different (II)

Name_____ Date_____

Circle the one that is **different**.

Unit 6

Visual Discrimination: Different (III)

Name_____ Date_____

Circle the one that is **different**.

Sizes: Same (1)

Name_____ Date_____

 same size

Circle the cows that are the **same size**.

Circle the sheep that are the **same size**.

Unit 6

Sizes: Same (II)

Name_____ Date_____

Color the ones that are the **same size**.

 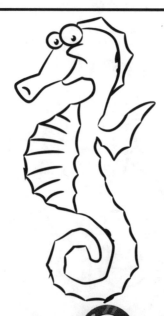 

Sizes: Tall

Name_____ Date_____

tall   **short**

Color the **tall** one.

  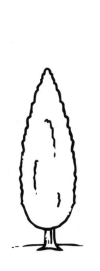 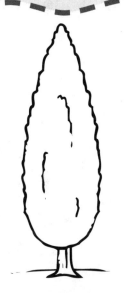

Unit 6

Premium Education Math: Preschool © Learning Horizons

Sizes: Short

Name_____ Date_____

short  tall

Color the **short** one.

Sizes: Big

Name_____ Date_____

big  little

Color the **big** one.

Sizes: Little

Name_____ Date_____

little     **big**

Color the **little** one.

Sizes: Long

Name_____ Date_____

long
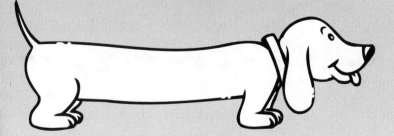

short

Color the **long** one.

Sizes: Short

Name_____ Date_____

short   **long**

Color the **short** one.

Sizes: Tallest

Name_____   Date_____

tallest  **shortest**

Circle the **tallest** one.

Unit 6

Sizes: Shortest

Name_____ Date_____

shortest    tallest

Circle the **shortest** one.

Sizes: Biggest

Name _____ Date _____

biggest   **littlest**

Color the **biggest** one.

Sizes: Littlest

Name_____ Date_____

littlest   **biggest**

Color the **littlest** one.

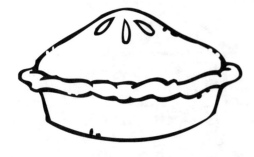  

Sizes: Longest

Name_____ Date_____

longest  **shortest**

Circle the **longest** one.

Sizes: Shortest

Name_____ Date_____

shortest   longest

Color the **shortest** one.

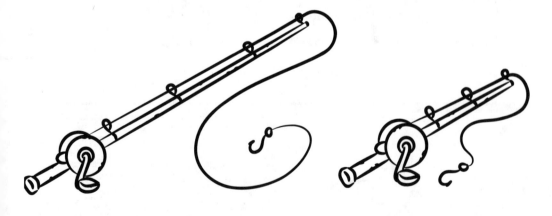 

Sizes: Review (1)

Name_____ Date_____

Color the big animal. Circle the little animal.

Sizes: Review (II)

Name_____ Date_____

Color the long one. Circle the short one.

Patterns: Shapes (1)

Name_____ Date_____

Color what comes next.

Unit 6

Teaching Tip: Have the child point to and name the shapes in each row while moving his or her finger from left to right. Help him or her decide which shape comes next. Then have the child color the shape that comes next.

Patterns: Shapes (II)

Name_____ Date_____

Color what comes next.

Teaching Tip: Have the child point to and name the shapes in each row while moving his or her finger from left to right. Help him or her decide which shape comes next. Then have the child color the shape that comes next.

Patterns: Shapes (III)

Name_____ Date_____

Draw a line to what comes next.

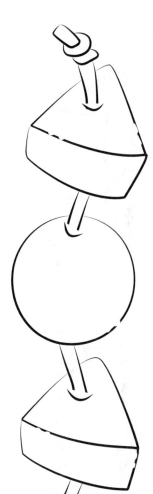

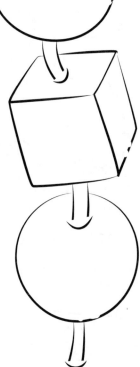

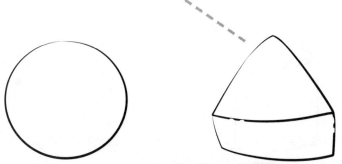

Patterns: Shapes (IV)

Name_____ Date_____

Draw a line to what comes next.

 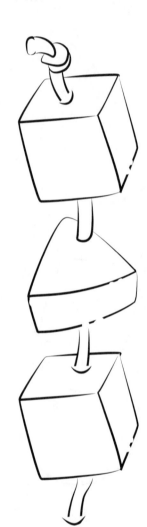 

  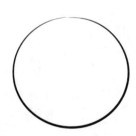

Patterns: Numbers (1)

Name_____ Date_____

Continue the number pattern.

1 2 3 4 5 ____

10 9 8 7 6 ____

4 5 6 7 8 ____

7 6 5 4 3 ____

Teaching Tip: Have the child point to and name the numbers in each row while moving his or her finger from left to right. Help him or her decide which number comes next. Then have the child write the number that comes next.

Patterns: Numbers (II)

Name_____ Date_____

Continue the number pattern.

1 2 1 2 1 ____

7 8 7 8 7 ____

3 5 3 5 3 ____

10 4 10 4 10 ____

Teaching Tip: Have the child point to and name the numbers in each row while moving his or her finger from left to right. Help him or her decide which number comes next. Then have the child write the number that comes next.

Time (1)

Name_____ Date_____

Color what **tells time**.

Unit 7

Time (II)

Name_____ Date_____

Draw an **X** on what **does not tell time**.

Time (III)

Name_____ Date_____

Color what **takes more time**.

Money (I)

Name_____ Date_____

Color what is **money**.

Money (II)

Name_____ Date_____

Draw an **X** on what is **not money**.

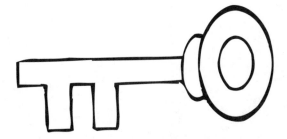

Money (III)

Name_____ Date_____

Color the group that shows **more money**.

Measurement (1)

Name_____ Date_____

Color what is **used to measure**.

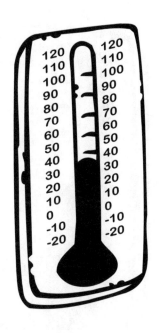
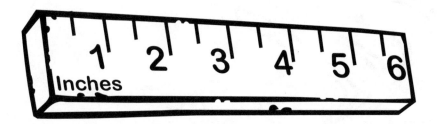

Measurement (II)

Name_____ Date_____

Draw an **X** on what is **not used to measure**.

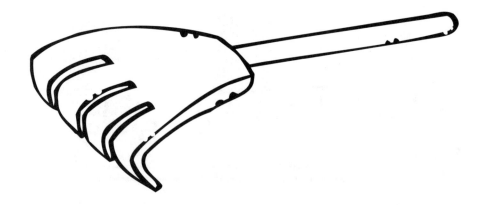
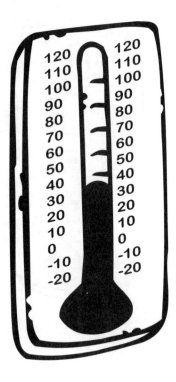
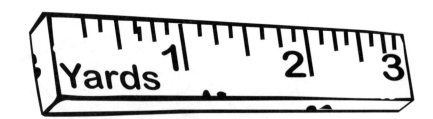

Practice Test: Numerals

Name_____ Date_____

Print the numerals.

1 2 3 4 5

6 7 8 9 10

Unit 8

Practice Test: Shapes

Name_____ Date_____

Fill in the circle next to the correct answer.

Which is a **circle**? Mark them all.

   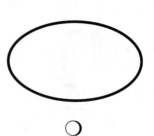 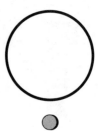

Which is a **square**? Mark them all.

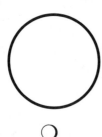  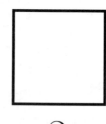 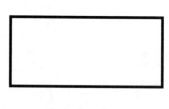

Which is a **triangle**? Mark them all.

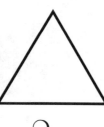  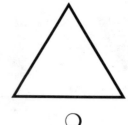 

Which is a **rectangle**? Mark them all.

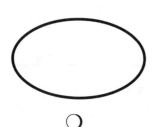  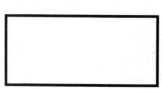 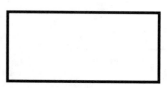

Premium Education Math: Preschool © Learning Horizons

Practice Test: Fine Motor Skills

Name_____ Date_____

Trace the shapes.

Trace the shapes to complete the picture.

Unit 8

Practice Test: Counting

Name_____ Date_____

Fill in the circle next to the correct answer. Count and show the correct number.

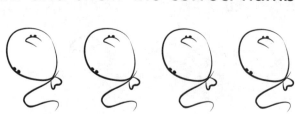

○ 3   ● 4   ○ 6

○ 1   ○ 6   ○ 3

○ 3   ○ 9   ○ 2

○ 10   ○ 7   ○ 2

○ 5   ○ 4   ○ 6

○ 1   ○ 6   ○ 5

○ 8   ○ 4   ○ 7

○ 4   ○ 5   ○ 2

○ 8   ○ 10   ○ 9

○ 10   ○ 2   ○ 1

Premium Education Math: Preschool © Learning Horizons

Practice Test: Sets

Name _____ Date _____

Fill in the circle next to the correct answer.

Choose the group with the **matching number**.

Choose the group with the **matching number**.

Which group has **more**?

○   ○

Which group has **less**?

○   ○

Practice Test: Visual Discrimination

Name_____ Date_____

Fill in the circle next to the correct answer.

Which ones are the **same**? Mark them all.

○   ○   ○   ○

Which ones are the **same**? Mark them all.

○   ○   ○   ○

Which one is **different**?

○   ○   ○   ○

Which one is **different**?

○   ○   ○   ○

Practice Test: Sizes and Patterns

Name_____ Date_____

Fill in the circle next to the correct answer.

Which is the **biggest**?

○ ○ ○

Which is the **longest**?

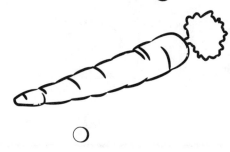  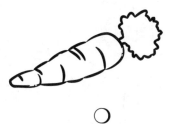
○ ○ ○

Which shape **comes next**?

 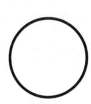  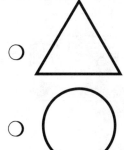

Which number **comes next**?

1 5 1 ____   ○ 1

○ 5

Unit 8

Practice Test: Measurement

Name_____ Date_____

Fill in the circle next to the correct answer.

Which tell **time**? Mark them all.

○ ○ ○ ○

Which are **money**? Mark them all.

○ ○ ○ ○

Which are used to **measure**? Mark them all.

○ ○ ○ ○

Which **does not belong**?

○ ○ ○

Answer Key

Page 3

Page 4

Page 5

Page 6

Page 7

Page 8

Page 9
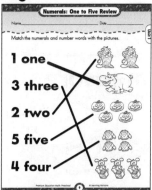

Page 10

Page 11

Page 12
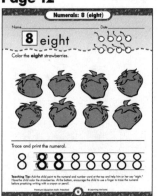

Page 13
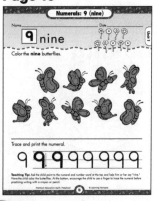

Page 14

Answer Key

Page 15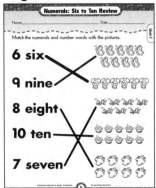

Page 16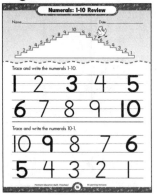

Page 17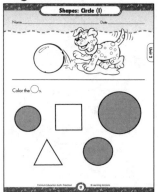

Page 18

Page 19

Page 20

Page 21

Page 22

Page 23

Page 24

Page 25

Page 26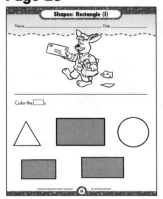

Page 27

Page 28

Page 29

Page 30

Premium Education Math: Preschool © Learning Horizons

Answer Key

Page 31

Page 32

Page 33

Page 34

Page 35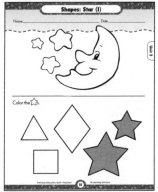

Page 36

Page 37

Page 38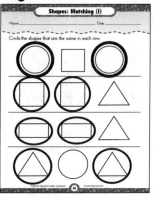

Page 39

Page 40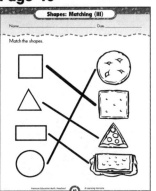

Page 41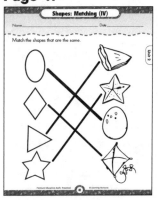

Page 42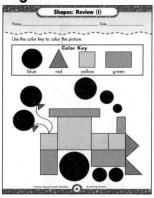

Page 43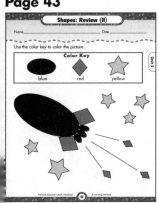

Page 44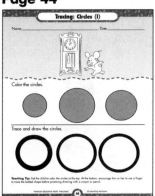

Page 45

Page 46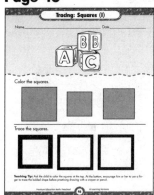

Answer Key

Page 47

Page 48
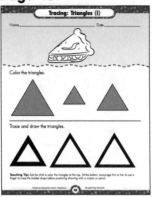

Page 49
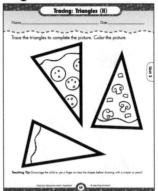

Page 50
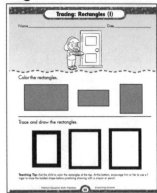

Page 51
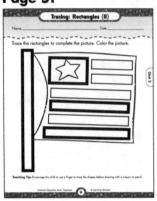

Page 52
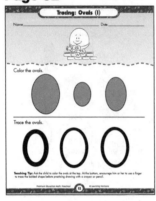

Page 53
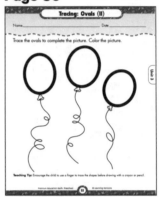

Page 54
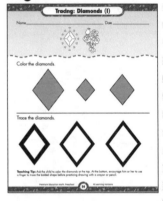

Page 55
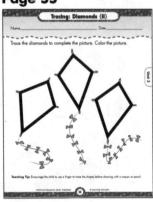

Page 56
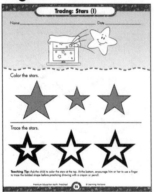

Page 57
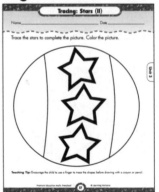

Page 58
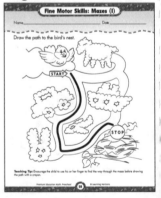

Page 59

Page 60

Page 61

Page 62

Answer Key

Page 63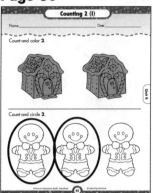

Page 64

Page 65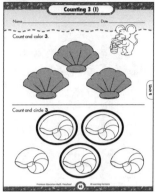

Page 66

Page 67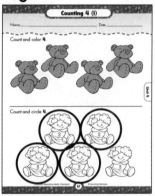

Page 68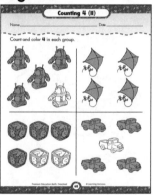

Page 69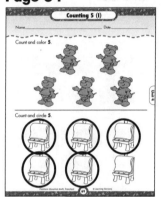

Page 70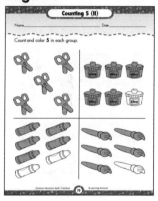

Page 71

Page 72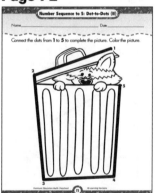

Page 73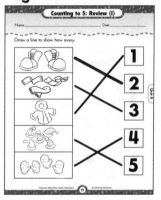

Page 74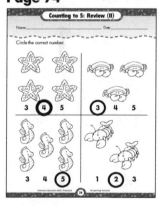

Page 75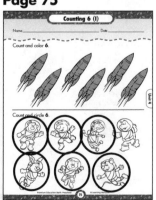

Page 76

Page 77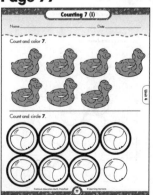

Page 78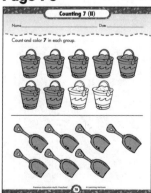

Premium Education Math: Preschool © Learning Horizons

Answer Key

Page 79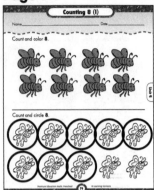

Page 80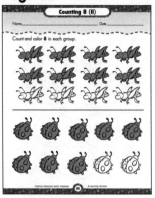

Page 81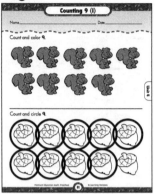

Page 82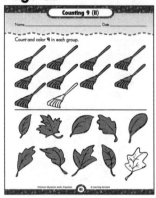

Page 83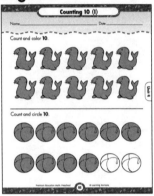

Page 84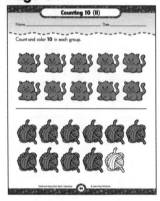

Page 85

Page 86

Page 87

Page 88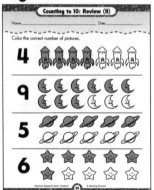

Page 89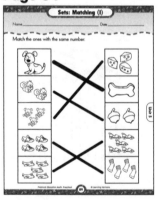

Page 90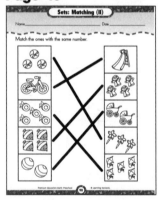

Page 91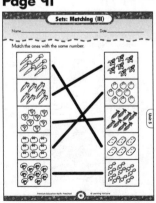

Page 92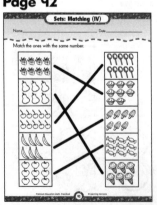

Page 93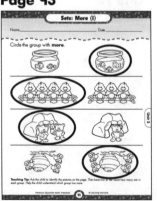

Page 94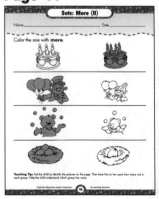

Answer Key

Page 95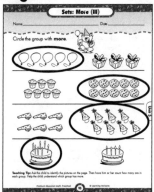
Page 96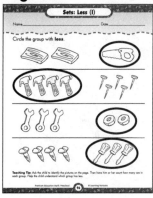
Page 97
Page 98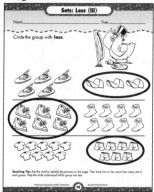

Page 99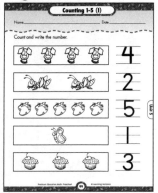
Page 100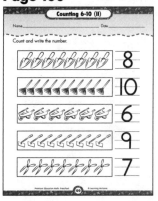
Page 101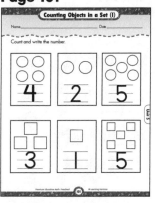
Page 102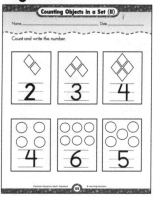

Page 103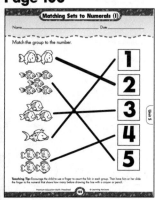
Page 104
Page 105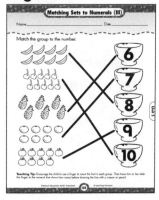
Page 106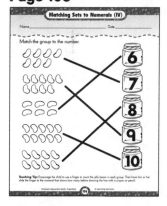

Page 107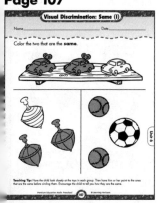
Page 108
Page 109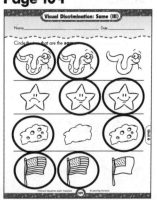
Page 110

Premium Education Math: Preschool © Learning Horizons

Answer Key

Page 111

Page 112

Page 113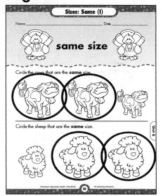

Page 114

Page 115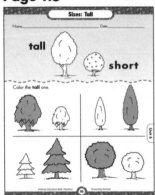

Page 116

Page 117

Page 118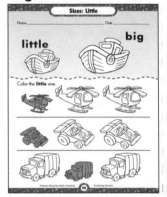

Page 119

Page 120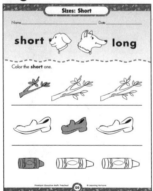

Page 121

Page 122

Page 123

Page 124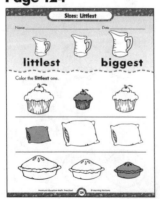

Page 125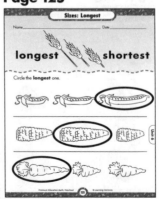

Page 126

Premium Education Math: Preschool © Learning Horizons

Answer Key

Page 127

Page 128
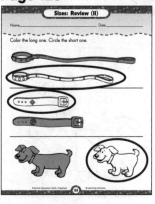

Page 129
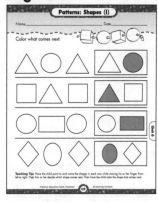

Page 130
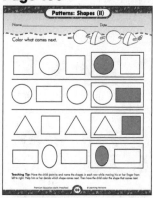

Page 131
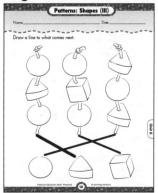

Page 132
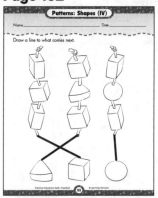

Page 133

Page 134
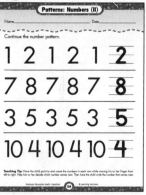

Page 135
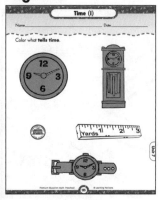

Page 136
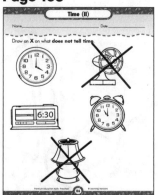

Page 137

Page 138

Page 139

Page 140
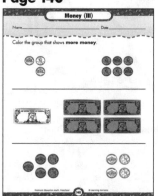

Page 141
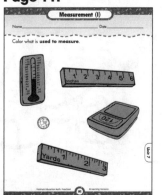

Page 142

Premium Education Math: Preschool © Learning Horizons

Answer Key

Page 143
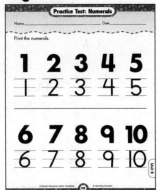

Page 144
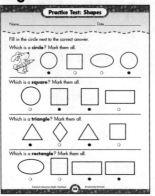

Page 145
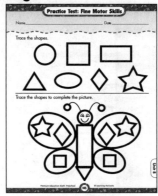

Page 146
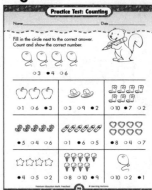

Page 147
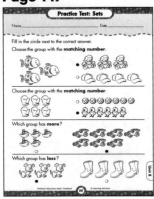

Page 148

Page 149
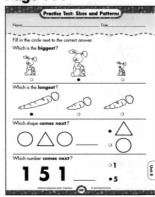

Page 150